VOLUNTEERS CAMP AND FIELD BOOK

CONTAINING

USEFUL AND GENERAL INFORMATION ON THE ART AND SCIENCE OF WAR, FOR THE LEISURE MOMENTS OF THE SOLDIER

1862 EDITION WITH NOTES AND AN APPENDIX

BY

JOHN P. CURRY

COMPILED AND EDITED BY

WILLIAM B. SARGEANT

AND

JOHN W. BRINSFIELD

MERCER UNIVERSITY PRESS
MACON, GEORGIA

MERCER
UNIVERSITY PRESS

Endowed by
TOM WATSON BROWN
and
THE WATSON-BROWN FOUNDATION, INC.

MUP/P397

1400 Coleman Avenue
Macon, Georgia 31207

First Edition.

ISBN 978-0-88146-169-5

Books published by Mercer University Press are printed on acid free paper that meets the requirements of American National Standard for Information Sciences—Permanence of Paper for Printed Library Materials.

Mercer University Press is a member of Green Press initiative (greenpressinitiative.org), a nonprofit organization working to help publishers and printers increase their use of recycled paper and decrease their use of fiber derived from endangered forests. This book is printed on recycled paper.

Book design by Burt&Burt.

Library of Congress Cataloging-in-Publication Data

PART FIRST
FIELD FORTIFICATIONS AND INTRENCHED POSITIONS

PART SECOND
ARTILLERY AND ARTILLERY PRACTICE

PART THIRD
HINTS ON SURGERY—ANTIDOTES FOR POISON & C.

PART FOURTH
CAVALRY AND CAVALRY MOVEMENTS

PART FIFTH
ORDER OF ENCAMPMENT FOR ARTILLERY, CAVALRY, AND INFANTRY, AND GENERAL DETAILS OF CAMP DUTY, COOKING & C.

PART SIXTH
ELEMENTARY PRINCIPLES OF THE MANUAL—FORMATION OF COMPANY AND REGIMENT

APPENDIX
THE LIFE AND WRITINGS OF JOHN PENN CURRY

VOLUNTEERS'

CAMP AND FIELD

BOOK

CONTAINING

USEFUL AND GENERAL INFORMATION ON THE ART AND SCIENCE OF WAR, FOR THE LEISURE MOMENTS OF THE SOLDIER.

BY JOHN P. CURRY.

RICHMOND:
WEST & JOHNSTON,
145 MAIN STREET,
1862.

"Original title page for Curry's Volunteers' Camp and Field Book."
Courtesy of William B. Sargeant.

PREFACE

THE BOOK

The Volunteers' Camp and Field Book

1861 AND 1862 EDITIONS

Like Daniel Emmett's minstrel song "Dixie," first performed in New York City in 1859, the *Volunteers' Camp and Field Book* by John Penn Curry was written in the North but adopted by the South during the Civil War. Even though there were a dozen soldiers' manuals already in print, the multitudes of volunteers reporting for duty in 1861 created a new demand for instruction books to help recruits learn military skills. Publishers in Boston, New York, Philadelphia, Cincinnati, and Columbus, Ohio, produced at least ten new or revised manuals for soldiers during the first year of the war.[1]

Setting Curry's book apart from many of the others was its appeal to soldiers in both the Union and Confederate armies. Curry's book was easy to read and contained practical tips on everything from field fortifications to preparing biscuits, slapjacks, and bread pudding. Curry advised soldiers to rub their socks with soap to prevent blisters while 'on the march.' Further, he suggested verification that a soldier was really

1 Francis A. Lord, *Army and Navy Textbooks and Manuals used by the North during the Civil War-Part II*, vol. 9 of *Company of Military Historians, Military Collector and Historian* (Carlisle, PA: U.S. Army Heritage and Education Center, 1957), 95-101.

dead before they buried him. His remedy for snakebite was "to drink whiskey to excess, or if you can procure the cedran-bean make a tea out of it and drink a half-pint every half-hour."[2]

Less than a year after Curry's book appeared in New York, the publishing firm of West and Johnston published a second, virtually identical edition in Richmond, Virginia.[3] The 1862 Southern edition was for the use of Confederate soldiers and their officers. There is no indication that Curry received any royalties from West and Johnston or that he even knew they had re-printed his book.

Curry's book may have been more popular in the South than in the North since competition for books of instruction was considerably less south of the Mason-Dixon line. Soldiers in the Army of Northern Virginia, particularly in General John B. Hood's Texas Brigade, valued Curry's comments on cooks and the preparation of coffee, bean soup, and meat.[4] Copies of Curry's book eventually showed up in North Carolina, Georgia, Alabama and Tennessee.[5]

John Penn Curry's sources and personal experiences are of interest for evaluating the historical importance of his book; especially since this handbook is one of the first published specifically for Civil War soldiers. For example, in the preface

---

2 John P. Curry, *Volunteers' Camp and Field Book* (New York: D. Appleton, 1861), 51. "Cedran-bean" may be a misprint in the original for cedarn, a reference to cedar tree seeds.

3 West and Johnston published a number of books during the Civil War including novels by female authors such as, Mary Elizabeth Braddon (1863) and Augusta Jane Evans (1864).

4 Lee Rainey, "Dining Out with Hood's Texans," *4th Texas Infantry*, http://members.aol .com/ h4texas/dining4.htm, accessed 2 November 2006.

to the *Volunteers' Camp and Field Book* and in a letter to Governor William Dennison of Ohio in 1861 Curry stated that he had been on campaigns and in the service of the U.S. government for many years before the Civil War began.[6] However, no U.S. Army service record for a John Penn Curry could be found. He was neither a graduate of the U.S. Military Academy at West Point nor of the U.S. Naval Academy at Annapolis. In addition, his name could not be matched to any of the soldiers named John Curry who served in the Mexican War. One wonders how Curry could have learned so much about the life of a soldier.

---

5 Original copies of the 1861 edition are very rare; however, one copy is located at each of the following university libraries in the South: Wake Forest University, the University of North Carolina at Chapel Hill, Duke University, and Emory University in Atlanta, Georgia. Microfilm copies are available in the libraries of Virginia Tech, University of Tennessee at Chatanooga, and Jacksonville State University in Jacksonville, Alabama.

6 Ohio Historical Society, "September 3, 1861," in the Civil War Documents Searchable Database, series 147-6: 187, http://ohiohistory.org/CivilWar GuideProject/. Note: John P. Curry did quote from the U.S. War Department, *Revised Regulations for the Army of the United States, 1861*, (New York: Harper and Brothers, 1861) 76, 79, 159, 163, 246. Curry mentioned "the old regulations" on page 22 of his book and then quoted or summarized parts of the 1861 *Regulations* on five other pages. Evidently Curry did not feel references to this source—or to any other source—were necessary.

THE AUTHOR
JOHN PENN CURRY, 1828–1909

If one could have interviewed John Penn Curry in New York City at the turn of the century, his stories might have captured the imagination of thousands of people, young and old, who loved tales of high adventure. Through the course of Curry's lifetime, which spanned more than 80 years, he panned for gold in California, fought Indians in the Yosemite Valley and later in Utah, and served on board a U.S. Navy warship in the Pacific, Arctic, and Atlantic Oceans. Add to this laundry list the fact that he wrote a book of instruction for Civil War soldiers, publicized the successes of Nevada silver mines as a "civil engineer" for potential investors,[7] worked as a newspaper correspondent in California and, then, a magazine editor in New York, patented an invention that made railroad travel more comfortable. He even once laid claim to prime Staten Island real estate which resulted in a law suit eventually heard by the United States Supreme Court.

Despite this long list of occupations and achievements, his life remains somewhat of a mystery. He wrote two books and scores of published articles; yet, he never wrote very much about himself. No evidence supports a marriage or children; instead he preferred to live in hotels and boarding houses. This lifestyle made it difficult to distinguish him from other similarly named men such as, a John Curry who lived in California

---

7 John Penn Curry, *Observations and Experiences among the Mineral Regions of Nevada* (New York: William Quinn, 1865) title page.

and in New York during the same time frame as John Penn Curry.

He outlived all of his immediate family members except for one younger sister and legal heir, Mrs. Emma Curry Riegel of Philadelphia. No obituary was written for Curry at his death; however, one friend described Curry as "a Christian gentleman" who was associated with Trinity Church (Episcopal) in New York City. Others thought that by the end of his life Curry had become a marginally dishonest opportunist who made money by unsettling land claims both in California and in New York. As far as is known, Curry never bothered to answer such vague charges, at least not in print. When Curry died in New York City in his 82nd year, his remains were taken to Philadelphia's Woodlands Cemetery for burial. There is no marker at his gravesite.

For those readers who want to know more about John Penn Curry, a chronology of his life with reference notes is included in the Appendix along with two articles regarding his experiences fighting Indians and renegades and with his visit to the Navigator Islands, now Samoa, as a naval officer in the 1850's. Curry described his own life as a "wandering, adventurous career"—clearly an understatement by a courageous and gifted writer.[8]

---

8 John Penn Curry, "An Adventure in the Wilds of California," *Gazlay's Pacific Monthly* (June 1865): 14.

THE DISCOVERY

It is not known how many copies of Curry's book were published in New York and Richmond during the Civil War, but by the end of the conflict individual copies were hard to find. In 1866 Alfred F. Smith, a school teacher in Randolph County, North Carolina, bought a copy of the *Volunteers' Camp and Field Book* at auction for $2.00, more than ten times its original cost of 15 cents per copy. Smith wrote an explanation inside the flyleaf: "Confederate book printed in Richmond during the rebellion."[9]

For 139 years Smith's descendants passed his books through four generations without knowing that Curry's volume was a historical treasure. In the summer of 2005, Smith's great-great grandson, William B. Sargeant, opened an old suitcase that had been used as a doorstop in his late mother's house. Inside the suitcase were several books published during the Civil War, Curry's *Volunteers' Camp and Field Book* among them.

At the urging of a relative who was a history teacher, Sargeant sent Curry's book to Mercer University Press for possible re-publication. The result is this reprint of the 1862 edition of the *Volunteers' Camp and Field Book*. In the interest of historical accuracy, the original spelling has been preserved except where noted.

---

9 Original in the possession of William B. Sargeant.

ACKNOWLEDGMENTS

I wish to thank my wife, Betsy Strout Sargeant, and my wonderful daughters, Mandy Sargeant, Joanna Sargeant Risley, and Katie Sargeant, for all of their help in putting this project together. I would like to thank my sister, Alice Ann Sargeant, for helping me with background information on how the book made its way from North Carolina to Florida. Thanks to Jim Herndon for his assistance in researching my family history, and to Karen Herndon for typing a copy of the original manuscript. This book is also dedicated to the memory of my mother, Anna Bostick Sargeant, the great-granddaughter of A. F. Smith. Special thanks to John W. Brinsfield, my wife's cousin, US Army Chaplain Corps historian and a professor of history. Without his influence and guidance this whole project would never have come together.

William Sargeant
Lakeland, Florida

"Recruits in Camp." Source:US Army Chaplain School Student Handout.

INTRODUCTORY.

The following pages, devoted to the art and science of war in camp and field, were written more for private instruction for the use of the members of several of the volunteer regiments now preparing for active service, than with any view at the outset for general publicity and circulation. The object of the writer was to give such hints and suggestions from his own experience in campaigning and camping out, as would contribute to the stock of information of those immediately under the command of friends who have had but little time to devote to the acquisition of military knowledge and few opportunities for practice on the subjects of which it treats. In so doing, he has endeavored to furnish the inexperienced volunteer, officer as well as private, with concise, yet comprehensive hand-book on every matter of interest connected with the employment of troops, in camp or on the field of action, whether in offensive or defensive operations; and such hints and suggestions that might not readily occur to them for alleviating much of the hardships and risks incidental to the profession of arms. Among the subjects touched upon in this little volume, will be found useful remarks on field fortifications, instructions on artillery practice and heavy ordinance, descriptive summary of explosive substances, cavalry movements, skirmishers and skirmishing, deployments, reconnoisances, outposts, surprises, and method of attack and defense. There will also be found ample suggestions on the art of firing at long and short range, protecting the body under close fire, cross-firing, details of duty, storming parties, order of battles, strategic movements and night attacks, manner of constructing rifle-pits and temporary barricades, together with an illustration of facing, marching, and wheeling movements, proper formation of the company and regiment, and the elementary principles of the manual of arms; to which are added, the order of encampments and a most useful compendium of camp-cooking, making earth ovens, and the treatment of wounds, poisons, &c., &c.

Part First

FIELD FORTIFICATIONS AND INTRENCHED POSITIONS.

"The Bayonet Charge: July 12, 1862." Source: Winslow Homer Civil War Prints

ATTACK AND DEFENSE.

The term field fortifications applies to such works of a temporary nature, that are constructed of earth, wood, bags of sand, etc., which may be erected during the operations of a campaign, to enable an armed force to resist with confidence and advantage the attack of one superior in numbers, discipline, or equipments. Permanent fortifications are those of a more permanent and durable character. Our sea-coast defences, built of brick or granite, and massive in their construction, come under that head, as are any positions, in fact, to be permanently occupied as military works. The meaning of the word intrenchment, or lines of intrenchments, signifies field works resorted to for the purpose of strengthening the position of an army, or detachment to be employed and occupied in defending the approaches to a stationary camp, or city, or to impede the progress of an invading enemy from penetrating within the line of defence or basis of operations. If woods, rivers, mountains, precipices, and rolling ground present themselves to the eye of an engineer intrusted with the erection of intrenchments, they are natural obstacles, which may be strengthened by military art, and are always considered in the plans to be adopted and marked out for the line of defence.

Intrenchments are hastily thrown up of such perishable materials as earth, wood, etc., afford. The chief object in laying them out is to enable the assailed to meet the enemy with success, by first compelling him to approach under every disadvantage of position, and then, when he is cut up by the fire of the works, and is exhausted by his efforts to carry them by vigorous assaults, with storming parties, to assume the offensive and drive him back at the point of the bayonet. This objective can only be attained by defending the works to the last extremity; and, unless successful, intrenchments would serve no other purpose than to shelter the assailed from the enemy's fire, without probably accomplishing a single military result.

The damage received by the enemy making the assault, from the fire of troops intrenched, who do not intend to make a vigorous defence, but can see safety only in retreat, would be necessarily trifling. But in encountering the courageous efforts of the defenders to hold their position, and repel the assault, the case would be far different, and the loss greater on the part of the assailants.

This principle leads to the rejection of all advanced works thrown up in front of the principal intrenchments, and to a succession of works one behind another, when the object is to afford retreat from the first to the last. Such manoeuvering as this only encourages the enemy, and demoralizes your own troops. If an advanced work is required for the defence of a point which cannot be defended by the principal line of intrenchments, it should have a reserve force in the immediate vicinity, and its garrison not permitted to retreat and abandon their positions until it is in absolute danger of being overpowered, when it should retire in good order, under cover of the fire of the reserve. As to retreating from one line to another, one of two things must happen: either the first line must be evacuated before the enemy enters, or crosses the ditch in front of your intrenchments, in good order, that the garrison may gain their second line in safety—in which case the first line will be of little service; or else, if the assailed wait until the assault is made on the parapet of the intrenchment before retreating, the enemy will enter pell-mell with them into the second line of defence which will then be of no possible service.

The men engaged in the immediate defence of intrenchments might be overpowered by the repeated assaults of the enemy, unless they were supported by a reserve.

The duties of a reserve are to make sorties on the flanks of an attacking column, charge the enemy at any critical moment of disorder, and to cover the retreat of the troops, if driven from the parapet. All intrenchments should contain a reserve proportioned to their importance.

The result of innumerable actions proves that the defence with the bayonet is the surest method of repelling an enemy. A sortie made upon the flank of the enemy, at the moment when his column is either checked by the fire of the works, or is in a state of confusion when crossing the ditch, will generally prove decisive in repelling his attack. For this reason—all intrenchments should be arranged with small outlets every fifty or one hundred yards, with a "flying bridge," to raise or throw across the ditch at such points, so as to facilitate sorties. This principle is only applicable, however, to engagements with large bodies of troops, defending works open in their rear. A small detachment should rest satisfied with repulsing the attack, and should not give up the advantage of their position, by sallying out to engage the enemy on equal terms. A retreat carries with it all the moral effects of a defeat; it inspirits the assailants, and renders the retreating corps timid, and impairs the confidence of the troops in their own self-reliance. Add to this the confusion that must ensue among the best-disciplined troops, under such circumstances, and the importance attached to the principle will be fully justified.

To enable the troops to fight with advantage, the intrenchments should shelter them from the enemy's fire.

In delivering his fire, a soldier usually aims directly to the front, so that the line of fire and the parapet make nearly a right angle with each other.

An attack is generally opened by a fire of the enemy's artillery, whose object is to silence the fire of the intrenchments, and to drive the assailed from the parapet. When they have accomplished this result, they throw out a storming party, which is composed usually of a detachment of engineer troops, a column of attack, and a reserve, who rush forward, under cover of the fire of their artillery, to the assault. The engineer troops precede the column of attack, and remove all obstacles that obstruct its passage to the ditch. This they endeavor to fill up with bags of earth, or cross over with pieces of wood. The bayonet should be resorted to by those in the intrenchments, as soon as the enemy succeeds in crossing the ditch.

The thickness of the parapet is regulated by the material used, the kind of attack, and its probable duration.

Shot will penetrate ordinary earth, when well rammed, the distances laid down as follows:

Musket ball, one foot six inches.

Six-pound shot, three and a-half to four and a-half feet.

Nine-pound shot, six and a-half to seven feet.

Twelve pound shot, eight and a-half to ten feet.

Eighteen and twenty-four pound shot, thirteen feet.

In order to insure the safety of the troops, these dimensions should be increased one-half, so that no shot shall penetrate more than two-thirds the entire distance.

The width of the ditch at the top should never be less than twelve feet. Its depth may depend upon circumstances. The earth thus thrown up makes the intrenchment, or parapet, more properly speaking. The earth should be well packed, and bundles of wood, stones, &c., interspersed here and there to strengthen it and prevent the earth from crumbling. Sods with the grass upwards are excellent for covering the parapet. On the inside of all earth works, there is a slope or shelf, made of earth, extending out a few feet, called the banquette, upon which the troops stand to fire over the parapet. The parapet should slope outwards, so as to enable the men to depress their pieces.

Every angle of defence should be ninety degrees, a line of defence should not exceed one hundred and sixty yards.

It should be remembered that a close fire of musketry is more deadly than one of artillery. The flanks of a line of intrenchments should be carefully guarded, or protected by a precipice, or river, or redoubts, water-courses or marshes.

A redoubt is an enclosed earth work, sometimes made square, others round, and with two or three pieces of artillery mounted in them.

A double redoubt, or star fort, is constructed as if two square redoubts crossed each other. The approaches to a line of intrenchments are frequently obstructed outside of the ditch by *trous de loup*, or pits, dug in the earth here and there, four and six feet deep, and the ground made as uneven as possible, by palisades, inundations, and mines.

Abatis works are made out of the large limbs of trees. The smaller branches are chopped off, and the ends, pointed and interlaced with care, are presented towards the enemy. Small trees, with the branches facing the enemy, and fastened horizontally on the ground outside of the ditch with stakes, are excellent obstacles.

The edge of the inside of the ditch is called the scarp, and the outside the counterscarp.

Fascines and gabions are bundles of wood and large wicker-work baskets made of willow and withes.[1]

The gabion revetment is seldom used, except for the trenches in the attack of permanent works, where it is desirable to place the troops speedily under cover from the enemy's grape or case-shot and musketry.

An enfilading fire is to sweep your flanks with artillery or musketry.

A plunging fire is from artillery, posted on a height, or rising piece of ground.

A ricochet shot applies more particularly to a shot fired on the water at a certain angle, so that it will skip, and strike two or three times in succession. It requires a skillful artillerist to fire one on the field, in land operations. They are terribly effective when fired at an advancing column.

---

1 A withe is a slender, flexible branch or twig used as a band or line to bind larger pieces of wood together.

Cross-firing is where the fires (or shot) cross each other, and is the most effective plan to adopt in working a battery.

Rifle pits are small holes dug in the ground, large enough to admit one rifleman. They are rarely needed, except in the attack on permanent works, or a long siege. Skilful shots are placed in them to pick off the enemy's artillerists.

They are easily and speedily constructed. Say a man is six foot tall, all that is necessarily for him to do, if he wishes to shield himself in a rifle pit, is to dig a hole three and half feet deep, and throw up the earth in front facing the enemy's position. In loose gravelly soil, he can use his bayonet for this purpose.

A shirt or two, fastened at the ends and filled with sand, will oftentimes be found a good safeguard from a musket or rifle ball, by the soldier screening himself behind it, from whence he can load and fire at pleasure, without any very great risk. It should also be remembered that it is somewhat difficult to sight a person, while lying on the ground, with any degree of accuracy of aim.

Sods cut with a common knife, and turned from the earth with the bayonet, also offer a shield from a bullet.

The approaches to a position may quickly be mined by sinking, a foot under the earth, old pork or beef-barrels, filled with combustibles. In the absence of a hose or fuse to stretch underground, or magnetic wires leading to your position, by which to fire the mine, a little gully or a dozen may be constructed from the mine to your post, charged with powder and all traces obliterated by covering with small sods.

The loss of life in battles, in the open field, is generally more bloody, though not always, than assaults of intrenched positions. The reason is, that in the former case the entire army is usually engaged, but in the latter, only the storming column. The loss of those actually occupied in the attack is always heavier, however, than the average loss of an army in the open field.

A storming column relies mostly on the bayonet for making a successful assault.

To storm an intrenched position, defended, by an equal force, should never be resorted to except in case of absolute military necessity. In such an emergency, the exercise of the greatest military sagacity on the part of the officer planning the attack is required, and the most determined valor of the assaulting party absolutely requisite. The chances of success even then are more than doubtful, unless aid is rendered them by a vigorous bombardment and cannonading on the flanks of the enemy's position, sorties and efforts made to withdraw a portion, if not all, of the defenders from the point of approach of the attacking column. Every device and stratagem must be made, by feigned attacks, to enable the assaulting column to attack at the proper moment and in the weakest place. Once inside the intrenched position, reserved columns should instantly be hurried forward to their succor.

RECONNOISSANCES.[2]

The term reconnaissance is a detailed description of any extent of country, with a view of ascertaining its resources for the movements and subsistence of troops. The importance of reliable reconnoissances for all military operations is so great that they should be thorough and complete. It consists of two parts: a map, describing the topographical features of the country, and a descriptive memoir.

The reconnaissance of an enemy's position is one of a special nature, and none but a skilful officer should be intrusted with it. He seeks information of the force of the enemy, the number of guns; and if intrenched, his exact position, the extent of ground he occupies and its character.

In troops marching, he will endeavor to obtain their number. To an unpractised eye, a column composed of five hundred men is often estimated at double the number.

Troops marching in close or open column may be computed by the average number in platoons, the distance apart, and the time occupied in passing a given point, whether by quick or double quick time. A stop watch is necessary, and the result may easily be ascertained.

RECONNOITERING PARTIES.

Reconnoitering parties, when making observations of the enemy, should observe the following precautions: to leave sentinels or small posts at intervals; to transmit intelligence to

---

2 Curry uses the old French spelling of reconnaissance, an exploration of enemy territory.

the advance posts of the army, unless the return is to be by a different route; to march with caution, beware of ambuscades, avoid fighting, and see if possible without being seen; to keep an advanced guard; to send well-mounted scouts ahead of the advanced guard, and on the flank of the party; to instruct the scouts that no two should enter a defile, or mount a hill together, but to go one at a time, while one watches to carry the news if the other is taken, and warn them.

INTRENCHED POSTS.

Unless an army be acting on the defensive, no post should be intrenched, except to cover the weak part of the lines, or at points which the enemy cannot avoid, or in mountain warfare, or to close a defile, or cover winterquarters.

Any intrenchment that requires artillery is considered a post, and a guard or garrison, and commander assigned to it. In dark weather he redoubles his vigilance. He permits no flags of truce, deserters, or strangers to enter. If a flag ought to pass his post, he bandages his eyes that he may not observe his position. He refuses admittance to a relief or other party until he has carefully examined them. In case of an attack, he does not wait for orders, but makes the best defence he is capable of. If unsuccessful, he may spike his guns and rejoin the army under cover of night.

DEPLOYMENTS.

A company may be deployed as skirmishers, either forward or by the flank. It is deployed forward, when it is behind the line on which it is to be established; and by the flank, when already on that line. In both cases, it is divided into three platoons; the right and left platoons, denominated, respectively, First and Second platoons, form the line of skirmishers; the centre platoon, the Reserve.[3] A line of skirmishers should, if possible, be alligned; but, to obtain this regularity, advantages which the ground may present for covering the men, ought not to be neglected.

The intervals between files of skirmishers should not exceed fifteen paces.

The captain of a company gives a general superintendence to the deployment of skirmishers, and then throws himself about seventy paces in rear of the line, taking with him the first bugler, or drummer, and two or three men from the reserve.

Skirmishers should be selected from among the best marksmen in the ranks.

They are thrown out after a line of battle is formed to open the engagement; also, on the flanks of an army in retreating or advancing, co-operating with the rear and vanguards of the same.

---

3 Capitalization as in the original printing.

GUERILLA WARFARE.

This, the most barbarous and inhuman mode of warfare known, and by no means recognized among honorable combatants, is generally resorted to by irregular troops for the purpose of harassing and annoying an invading army entering an enemy's territory. They usually murder for pay and plunder, and are not prompted by any spirit of patriotism or honor. It is seldom they wear any uniform or distinguishing mark by which they may be known from peaceful persons, but assume such disguises as may best answer their purpose, such as farmers, laboring men and drovers, so that they may more readily escape detection, by throwing their arms into the nearest ditch in the event they are hard pressed. If captured with arms in their hands, they justly merit hanging by the roadside; and as no quarter or mercy is recognized by the guerilla, none should be shown him. If taken, even without arms, it is customary for them to affect the character of non-combatants, and hope by imposing upon the credulity of their captors by various pretences to be allowed to run at large again, to pursue the same murderous mode of warfare. The greatest care and precaution, however, should be adopted, and the most rigid scrutiny resorted to by the captors in detecting such rascals, so that really innocent non-combatants may not suffer for the guilty.

Guerilla warfare consists, mainly, in making the night attacks, waylaying strangers, the free use of poison, incendiarism and vandalism, firing upon victims from ambush, under cover of hedges and ditches, and in robbery, pillage, and assassination.

Other infamous means are oftentimes resorted to by them, such as poisoning wells and streams on the route of march; furnishing poisoned food to a scouting party, or an army, under the cloak of friendship and fair dealing, and in massacring the wounded left after a route on the field of battle.

If a guerilla is caught, no clemency whatever should be extended to him.

Light troops and mounted rangers are the most effective in suppressing guerilla bands; and make excellent flankers when a column is advancing or in retreat.

FUGITIVE-WANDERINGS.

The chief fault of recruits is in firing too high; but this must not be understood that they should fire too low by any means. Only at point-blank range should the piece be aimed directly at the object. If beyond that distance, the muzzle of the piece should be slightly elevated, so as to allow for the falling of the ball by its own weight. In a long flight this always occurs.

As an illustration: say, at one thousand yards' distance you wish to hit an adversary in the breast, you must aim for his head; and again, if the distance off is still greater than that, you must aim a little higher. Care, however, must be taken to make a *line* shot of it, and not aim to the right or left of the object, but directly over it. If a strong wind is blowing from the *left* of you, aim a little to the *right* of your object; as, if you aimed directly at it, the force of the wind would carry your ball out of range. If the wind is blowing on your right side, then fire to the left of the object aimed at.

When firing, the least motion of your arms or body should be carefully avoided, as the slightest movement on the instant of pulling trigger will destroy the aim, and throw the ball out of line.

Inside of point-blank range, the muzzle of the piece should *fall half an inch below the right eye* —- that is, if your adversary stands on as level a piece of ground as yourself. If not, then aim directly at the centre of his breast.

Cross-firing is the most destructive. When a regiment is drawn up in line of battle, preparing for action, a piece of artillery posted at each flank can be made immensely serviceable in either attack or defence. If a column of the enemy approach, the pieces on the right and left flank should converge their fire directly on the front centre of the enemy's column, while those in the ranks fire direct. By thinning out his centre, his flanks must fall in to fill up the vacuum. A regiment should also be instructed in forming V's and W's, when a terrible cross-fire may be obtained by those on the left flank firing left oblique, and those stationed on the right, to the right oblique, while the artillery stationed at the *pivot* fires direct. It is almost impossible to penetrate this fire.

USEFUL HINTS AND EXPERIENCES.

There should be for every 100 men, as useful articles of camp furniture, ten camp kettles, in nests of five, to hold not less than four gallons, and made of Russia iron; also, ten mess pans of same capacity. Each man should have one tin plate, tin cup, knife, fork, and spoon, bearing his own mark. One cook and two assistants are enough for a company, where bread or biscuit is supplied. They make the coffee, bean soup, and cook the meat. Yeast powders should be furnished by the company commander, as it is not an article of the ration. Sugar and coffee rations have been increased by the army since the old regulations. Six tents for 100 men are enough, where the Sibley tent is used; one for every five men of the common army tent. Each man should have one heavy "four-point" blanket, which will answer for two on a march. Every man should have one light navy-flannel sack-coat, and pegged shoes, wide, well-nailed and large; three pair cotton socks, and each a soft felt hat, of a dark color—dark drab is preferable. Light or red caps should never be worn in action, for they offer a sure mark for an expert corps of riflemen. In a night attack on the enemy, or to effect a surprise, dark colors alone should be worn. When on such expeditions, some precaution must be observed not to expose burnished arms to the reflection of a light, or, if it can be avoided, even to the moon's rays. The best plan is to carry your arms reversed. Keep clear of the main roads, and take to the fields, or side trails, as you approach your point of surprise and attack. Move along the sides of hills, instead of crossing on top, and follow in the shade of trees when the moon is shining, and do not expose yourselves with the horizon in your rear, so that your forms may loom out distinctly against it—for sentinels are supposed to be, in time of war, ever on the alert.

Wash the body at least twice a week, and the face, chest, and arms daily.

If in camp, exercise daily by running, jumping, foot-ball or shinty.[4]

Keep the bowels free, and avoid eating oily meats, fat, or drinking an excess of coffee. The mind and body both become dull and sluggish by too much indulgence in the latter.

When overheated, and exposed to the sun, you may cool off by sopping the wrists with cold water and immersing the arms to the elbow. This cools the arteries of the arms, carrying the blood to the heart, and a few seconds after, the entire body feels refreshed. Take but a mouthful of water every few seconds, if you wish to quench your thirst. When overheated, drink then by what we may call sipping it carefully, lest you produce too sudden a reaction in the system, and endanger your life.

The head may be protected from the heat of the sun, when on a march, by laying a small piece of wet sponge directly over the crown, and keeping the head covered with your hat or cap.

If you have no sponge, your towel folded and wet, or some wet leaves, will answer the same purpose.

When firing at an enemy, the eyes should always be carefully shaded.

Provide yourself also with an India-rubber slip to lie on.

---

4 Shinty was a variation of the game of field hockey played with a curved slick and a ball or block of wood.

TO MAKE A TENT COMFORTABLE

The best position for a tent to be pitched, is on sloping ground; the opening facing the slope. If it is a single tent, without a "fly," the top may be covered with small branches of trees, attached to the tent pole. When you see a rain storm coming on, wet your tent and spread the water over the surface with your hands. This prevents the rain from striking through. Intrench your tent by turning up the sod all around it, and plastering the interstices between the sod with mud. It should be about eight inches high. If you wish to raise the interior of your tent a foot higher than the outside ground, sod it with sods. If you desire to keep it dry and warm, dig a hole in the centre a foot deep, with a small gully or ditch from one side of it, extending to the outside of the tent. Build a fire in the hole, and cover it and the gully with sod carefully fitted in: cut a small vent in the sod, covering the hole and another at the extreme end of the gully on the outside. These will create a draft, the vent outside carry off the smoke, the fire kept burning slowly, and the interior of the tent kept perfectly dry and comfortable.

TO MAKE AN EARTH OVEN.

This is made entirely of sod, build up around an oblong trench in the ground in which the fire is built. Pieces of hoop iron, or iron rammers, are placed over the trench to rest the bread, meat, and stews to be baked or cooked. The sides and top of the oven should be carefully covered, the sod on top fastened in their place by pieces of wood imbedded in them, and the interstices carefully closed with mud, mixed with straw or grass. A chimney about a foot high may also be constructed of sod; and a small vent, dug into the trench in front under the bars of iron, will afford sufficient draft for cooking purposes. The door leading to the oven, may either be a sod, or made as suits the pleasure of the soldier.

The whole can be constructed in fifteen minutes.

FULMINATING CAP POWDER.

Eight and a half parts of powder, fired with percussion caps, are quite equal in force to ten parts of gunpowder, fired in the ordinary way by means of pan and flint.

CONFINED AIR.

Air in the immediate juxtaposition of combustible material creates terrific force when fired, and is likely to explode the piece. Always be careful then that your cartridge is rammed well home, before discharging your piece.

TO PROTECT THE BODY UNDER CLOSE FIRE.

When required to come under close fire, the body may partially be protected in front from the chin to the thighs, by folding your blanket in five or six thicknesses, and fastening it under your cross-belts and waist-belts, leaving the end hanging over your stomach. If the blanket is wet, it affords more resistance.

Part Second.

ARTILLERY AND ARTILLERY PRACTICE

"Tossing a live artillery shell back across a trench, 1864 (Confederate sergeant) near Marietta, GA, June 1864" Source: J. William Jones, *Christ in the Camp,* 1904, 613.

MUNITIONS OF WAR AND EXPLOSIVE SUBSTANCES.

The wonderful improvements made in heavy ordnance and field artillery, within the past thirty years, have caused almost a complete and radical revolution in the old established system of warfare. The six, nine, twelve, eighteen, and twenty-four pound carronades, together with the long twelve and eighteen-pounders that were once considered such formidable weapons, are now spoken of by artillerists with feelings of unmitigated contempt when contrasted with eight, ten, and thirteen-inch chamber guns, rifled cannon, and the modern howitzer. But a few years since it seems when all our forts and defensive positions, as well as our men-of-war, were armed with the former guns. Nor were we alone in this position, for all the nations of Europe were similarly situated. To boast of a battery of twenty-four pound carronades, with a few long eighteens, on the decks of an old fashioned line-of-battle ship, or defiantly peering out from the casemates of one of our fortresses, was to consider ourselves well guaranteed for offensive and defensive demonstrations whenever it became necessary to uphold the dignity of the flag and maintain ourselves among the nations of the earth as a first-class military power, ready at all times to display our strength when occasion required. All our naval battles were fought with this calibre of guns, and yet we hear complaints that our ships carried heavier mettle in several of the actions afloat when we were pitted against the enemy than they did, although, as far as the number of pieces worked in battery on either side were concerned, in many of our naval engagements, they were nearly if not quite equal. They urged, in extenuation of the loss on their part, that we threw a heavier weight of mettle at each broadside than the calibre of their guns would admit, though on several occasions our forces had a less number of guns in action than they themselves. This only goes to show that, in our early history as a nation, England and France were no better off in this respect than we were. They had not then arrived at that perfection in inventing and constructing those terrible engines of destruction for which they are now so famous. During the Revolution, and in the war of 1812, no heavier mettle than six, four, and nine-pounders were used in land operations, except in sieges, when eighteen and twenty-four pounders were the largest cannon brought into requisition. The former came under the classification of light artillery, used on the field; and the latter, siege guns.

By what is known as chamber guns, one of the modern improvements in ordinace, is that the piece is larger in the interior space at the breech than at the bore. This space gives greater force to the explosion of the powder when it is fired, and of course must project the ball to further distance and with more velocity than the old models. They may be used to advantage in throwing shells, besides grape, shrapnel, canister, and round and double-beaded shot. In throwing shells, greater accuracy of aim and length of range are obtained than with the common mortar.

All fire-arms of large caliber come under the general term of artillery. The term ordnance applies mostly to heavy guns; but, at the same time, all guns in our service come under the general appellation of ordnance, whether light or heavy. The latter are commonly manufactured of iron, some cast, wrought; and others out of what is known as gun metal, composed of tin, iron and copper well fused, though other metals are oftentimes used in connection with iron and tin for the same purpose. Wrought iron guns are far superior to cast iron ones, for the reason that they have fewer flaws in them then the latter description, and are certainly less liable to burst than the others. Large wrought iron guns, however, should be welded in pieces, and only a small amount of heat employed.

Rifled cannon and howitzers—as light artillery—are manufactured more frequently out of steel and brass, though some are made out of the composition above mentioned.

HOWITZERS.

A howitzer is also a chambered gun, used principally for throwing shells, grape, and canister. They are of three kinds: the mountain howitzer, for carrying over rugged countries, lashed to mules, if necessary, and is quite light and short in pattern; the field and boat howitzer, of medium size, and the heavy, long sea-coast howitzer, for permanent works. The mountain howitzer carries usually six-pound projectiles; and occasionally we find one carrying twelve pounds. The former, being lighter, are preferable for mountain warfare. The field and boat howitzers are usually twelve pounders. The sea coast howitzer carries much heavier weight of metal, and is particularly adapted for hurling shells.

PAIXAN GUNS.

The Paixan gun is entirely a French invention, named after its originator, who was a general in the French army in 1822. He was the first person to bring into use shell guns, after which they were adopted by the American Government; the old carronades done away with for active service; the long twelve, eighteen, and twenty-four pounders cast aside, and our men of war and seaboard forts armed with medium sized and long thirty-twos and forty-twos of the Paixan make.

They are chamber guns, and continue to be in very general use.

To make ourselves better understood, we would mention that the range of a medium sized gun is not so great as a long piece.

COLUMBIADS.

The modern columbiad is nothing more than a slight improvement on the Paixan gun. They are much larger, however, heavier, and longer, and carry eight, ten, and eleven-inch projectiles, which means that the shot or shell is of the same width as the diameter of the bore of the gun. It is usual, moreover, in speaking of very large ordnance, instead of saying sixty-eight, ninety-two, and one hundred and twenty pounders, to refer to them as so many inch guns, whatever the diameter of the bore or muzzle may be.

The columbiad is a most formidable weapon when properly handled by experienced artillerists, and will carry heavier shot at longer reach than the ordinary Paixan, and still more so than that of the sea coast howitzer.

DALGHREN GUNS.

The Dalghren gun is named after Captain Dalghren, United States Navy, its inventor, and a gentleman who may very properly be considered one of the few most experienced gunner in either branch of the service. His thorough knowledge of the power and force of various explosive substances, and of ordnance, on which duty for many years he has been assigned, with great credit to himself and honor to the Government, places him in the front rank of our most distinguished officers. After undergoing a variety of dangerous experiments, in which his life was constantly in jeopardy, and with the exercise of considerable ingenuity and skill, he succeeded in perfecting his design so well that he produced a gun which many men of military experience consider far superior to any large ordnance yet introduced to the notice of the general public.

The peculiarity of the construction of the Dalghren gun is such that the danger of its bursting, except by an excessive overcharge, is almost a matter of impossibility, while the effectiveness and range of its fire are quite equal to the best of other patterns, without their objections.

As all guns burst at or near the breech, this one is made of enormous thickness at that part to what other cannon are, and gradually tapers smoothly down to the muzzle. The middle of the piece, which offers the least resistance to the explosion of a and laid on the trestle in rear of the arms. The sutler's stores are closed, and the kitchen fires put out.

ROUNDS AND PATROL.

Rounds and patrols are made at various times of the night by the officer and non-commissioned officers of the guard, to satisfy themselves of the vigilance of the police guard and sentinels.

AT REVEILLE.

Reveille is usually beaten at daybreak, when the police guard takes arms, the officer of the guard inspects it and the advanced post, and the sergeant replants the colors in place.

DOUBLE CHAIN OF SENTINELS.

When necessary, small outposts are stationed about the camp, forming what is called a double chain of sentinels. These posts are under the orders of the commander of the police force.

CAMPING PARTY.

When the camping party precedes the regiment, and the new police guard marches with the camping party, the guard, on reaching the camp, forms in line, thirty paces in front of the centre of the ground marked for the regiment. The officer of the guard furnishes the sentinels required.

PICKET OF A REGIMENT.

The picket of a regiment is a lieutenant, two sergeants, four corporals, a drummer, and about forty privates.

GRAND GUARDS.

Grand guards are the advanced posts of a camp, or cantonment, and are composed of cavalry and infantry, the cavalry furnishing the advanced sentinels. It is usually commanded by a captain. If there is no pass to be observed or defended, the guard is placed near the centre of the ground they are to observe.

In broken or mountainous countries, where the inhabitants are hostile, intermediate posts should be established, if the grand guard is distant from the camp, to keep up communication with the main body. Grand guards watch the enemy in front. The camp must furnish additional guards to protect their rear and secure their retreat, when attacked.

VIDETTES.

Videttes are mounted patrols. Their duty is to watch the enemy, and should endeavor in turn not to be seen. Videttes carry their pistols or carbines in their hands.

"Union Army Wagon Train, 1864." Source: Winslow Homer Civil War Prints.

A FLYING SENTINEL.

A flying sentinel is one who moves about on dark or stormy nights, examining the paths and hollows for an enemy.

CAVALRY PATROLS.

The duty of cavalry patrols is to examine a hostile country to a greater distance than infantry, and make known to the infantry guard all they observe. When patrols are sent beyond the advanced posts, the posts and sentinels should be informed.

CAMP FIRES.

There should be only one camp fire to each platoon. The fires of grand guards should be hid from view. At outposts liable to surprise, fires should not be permitted.

BEARERS OF FLAGS.

Bearers of flags are not permitted to pass the outer chain of sentinels. Their faces should be turned from the post, or army, and if necessary, their eyes bandaged, that they may not witness the number and force of their adversaries.[8]

APPOINTMENT OF ADJUTANTS.

Commanders of regiments appoint their own adjutants and non-commissioned staff of the regiment.

MARKED AND NUMBERED.

Every article except arms and accoutrements belonging to a regiment, should be marked with the name and number of the regiment.

Such articles belonging to companies, are to be marked with the letter of the company, in addition to the name and number of the regiment.

---

8 Curry is referring to enemy personnel who come into camp under a flag of truce.

INSPECTION OF KITCHEN.

In camps or barracks, company officers should inspect the kitchen daily, as to economy in cooking and cleanliness of the utensils.

MESS FURNITURE.

The only mess furniture of the soldier, is one tin plate, tin cup, knife, fork and spoon, to be carried by himself on the march.

COMPANY BOOKS.

The books allowed to each company: one descriptive book, one clothing book, one order book, and one morning report book.

LAUNDRESSES.

Four women are allowed to each company as washer-women, and receive one ration per day each.

ARRESTS.

None but commanding officers have power to place officers under arrest.

All prisoners without written charges preferred against them, will be released by the officer of the day at guard-mounting, unless the commanding cfficer orders their detention.

MILITARY HONORS AND SALUTES.

In saluting a general officer, it should be done by cavalry with sabres presented, trumpets sounding the march, all the officers saluting and standards dropping. By infantry with drums beating march, arms presented, colors dropping.

The President is to be saluted with the highest honors.

Guards turn out and present arms to general officers as often as they pass them.

To commanders of regiments, garrison or camp, their own guard turn out, and present arms once a day, after which they turn out with shouldered arms.

Officers when they pass each other should salute by touching the hat; inferiors always salute superiors first, which the latter should promptly return.

FUNERAL HONORS.

The funeral escort of a Commander-in-Chief consists of a regiment of infantry, squadron of cavalry, and battery of six pieces of artillery.

Major-General.—A regiment of infantry, squadron of cavalry, and four pieces of artillery.

Brigadier-General.—A regiment of infantry, troop of cavalry, and two pieces of artillery.

For a Colonel.—A regiment.

Lieutenant-Colonel.—Six companies.

Major.—Four companies.

Captain.—One company.

Subaltern.—Half a company.

The escort should always be commanded by an officer of the same rank with the deceased. If none be present, then the next inferior rank.

That of a Sergeant.—Of fourteen rank and file, commanded by a Sergeant.

Corporal.—Of twelve rank and file.

Private.—Of eight rank and file commanded by a Corporal. For officers three volleys are fired over their graves.

PRISONERS OF WAR.

A report of prisoners, and of the number and description of the killed and wounded of the enemy, should be forwarded to the adjutant-general's office.

ARRANGEMENT OF TROOPS.

Arrangement of troops on parade and in order of battle, is; 1st, light infantry; 2d, infantry of the line; 3d, light cavalry, 4th, cavalry of the line; 5th, heavy cavalry; artillery and engineers are in the centre of the brigades or corps to which they are attached. Volunteers and militia take position on the left of the regular troops of the same arm.

MARCHES.

Cavalry and infantry do not march together, unless the proximity of the enemy makes it necessary. Batteries of artillery and their caissons move with the corps to which they are attached. The field train and ambulances march in rear of the column, and the baggage with the rear guard. When necessary, the orders specify the rations the men are to carry in their haversacks. The execution of marching orders should not be delayed. If the commander is not at the head of his troops when they are to march, the next in rank puts the column in motion. On the march no one shall fire a gun or cry *halt* or *march* without orders. Soldiers are not to stop for water, the canteens should be filled before starting. Halts to rest and reform the troops are frequent during the day, depending on the object and length of the march. No honors are paid by troops on the march, or at halts. The sick march with the wagons.

ORDER OF BATTLES.

At proper distances from the enemy, the troops are formed for the attack in several lines. If only two can be formed, some battalions in column are placed behind the wing of the second line. The lines may be formed of troops in column or in order of battle according to the ground and plan of attack. The reserve is formed of the best troops of foot and horse, to complete a victory, or make a good retreat. It is placed in the rear of the centre, or chief point of attack or defence. The cavalry should be distributed on the wings and at the centre, on favorable ground. In the attack, the artillery is employed to silence the enemy's batteries in position. In the defence, it is to direct its fire on advancing troops. In either case the fire of artillery should be concentrated. In battles and military operations, it is better to assume the offensive, and put the enemy on the defensive if possible; but to be safe in doing so requires a larger force than the enemy, or more favorable ground. If the attack of an enemy is repulsed, offensive operations against him must at once be taken. Soldiers must not leave the ranks to strip, or rob the dead, unless permission is given. The ambulance depot is generally established at some building nearest the scene of battle. A *red flag* is flying over it, to guide the ambulance wagons, and the wounded who can walk to it. After an action the munitions of war left on the field, are collected by the ordinance officers. Prisoners of war are disarmed and sent to the rear. The provost-marshal takes charge of prisoners.

"Thanksgiving in Camp at the Sutler's Store" Source: Winslow Homer Civil War Prints,

QUARTERMASTER'S DEPARTMENT.

This department provides the quarters and transportation of the army, storage and transportation for all army supplies, clothing, camp and garrison equipage, cavalry and artillery horses, fuel, forage, straw, and stationery. He should take triplicate receipts for all his expenditures, and account for all disbursements made by him, &c., to head-quarters. He pays, also, all the incidental expenses of the army, of the pursuit and detection of deserters, burials of officers and soldiers, of spies and guides, medicines, &c., &c.

SUBSISTENCE DEPARTMENT.

The commissary attends to the subsistence of troops. The rations in the regular army are as follows: Three quarters of a pound of pork or bacon, or one and a quarter pounds of fresh or salt beef; eighteen ounces of bread or flour, or twelve ounces of hard bread, or one and a quarter pounds corn meal; and at the rate, to one hundred rations, of eight quarts of peas or beans, or in lieu thereof, ten pounds of rice; six pounds coffee; twelve pounds sugar; four quarts of vinegar; one and a half pounds of tallow; one pound of sperm candles: four pounds of soap; and two quarts of salt. Sometimes molasses, picked onions, sour krout, fresh vegetable, dried apples or peaches, are substituted for some of the regular rations.

MEDICAL DEPARTMENT.

This department is devoted exclusively to the sick and wounded of the army.

PAY DEPARTMENT.

This department attends to the pay and accounts of the officers and men.

DESERTERS.

Deserters are disarmed at the advanced posts, and sent to the commanding officer of the guard, who questions them concerning the weakness of the enemy, the position or strength of his forces. If many come in at one time, they should be received cautiously, and only a few together, the guard standing to arms meanwhile, within a few paces. They are sent in the morning to the officer of the day, who conducts them before the colonel, or if attached to a brigade, to the brigade-general.

TO SALUTE WITH THE SWORD OR SABRE.

At the distance of six paces from the person to be saluted, raise the sword or sabre perpendicularly with the right hand, the point up, the flat of the blade opposite to the right eye, the guard at the height of the shoulder, and the elbow supported on the body. Drop the point of the sword or sabre in extending the arm, so that the right hand may be brought to the right thigh, and remain in that position until the person to whom the salute is rendered shall be passed, or shall have passed six paces.

COLOR SALUTES.

In ranks the color-bearer salutes by gliding the right hand along the lance to the height of the eye, the heel of the lance remaining at the hip. Lower the lance by extending the right arm.

INSPECTION OF TROOPS.

The inspections of troops are generally preceded by a review. The inspecting officer requires every man to present a cleanly and soldierly appearance.

SURGEON'S CALL.

When the surgeon's call is sounded, the sick able to go out, are conducted by the first sergeants of companies to the hospital.

DISCIPLINE.

To obtain the necessary amount of skill in firing a gun, or working a battery, a vast deal of exercising is necessary, alike for the officers to direct and the men to execute the movements. A man must rely entirely for his safety upon the skill and promptitude of his movements for a successful issue. The multifarious and varied contingencies of actual service require that strict discipline should be had at all times.

A FLYING COLUMN.

A flying column is composed of light cavalry, light artillery, and light infantry or rifles, or either of them. They effect strategic movements, lay in ambuscades, cut off supplies, and take prisoners after a rout.

IN FIRING.

Great practice and judgment alone are required to do so accurately. The inclination of the ground should always be taken into consideration.

PRECEDENCE OF REGIMENTS AND CORPS.

The precedence of regiments and corps is as follows: 1st, light artillery; 2d, light dragoons; 3d, other regular cavalry; 4th, artillery; 5th, infantry; 6th, marine corps, 7th, riflemen; 8th; volunteer corps; 9th, the militia.

This order of precedence refers to parades. On all other occasions, the several regiments and corps are distributed as the commanding officer may judge best adapted to the service they are on.

DAILY DETAILS AND DUTIES.

The daily duty is announced in orders. In large camps there is a general officer of the day for each division; a field officer of the day for each brigade; a captain of the day for each regiment; and such general and regimental staff officers of the day as may be necessary to attend to various details, and to receive and execute orders, according to their respective stations. In every regiment and garrison, besides the officer of the day, there should be detached daily, if the strength of the garrison will permit, a subaltern, four non-commissioned officers, a drummer, and such fatigue parties as circumstances may require.

The general of the day superintends the regularity and discipline of the camp. The field officer of the day superintends the camp of the brigade. The captain of the day, the cleanliness and regularity of the camp, or quarters of the regiment.

GUARDS.

The principal guards are:—1. Outposts and picket-guards. 2. Camp and garrison guards. 3. General officer's guard.

TRIMMINGS AND COLORS.

The color of the trimmings to the uniforms, as adopted for the various corps of the army, is as follows:

For artillerymen—scarlet.

Infantry—light, or sky-blue.

Riflemen—emerald-green, or medium.

Dragoons—orange.

Aides-de-camp—buff.

By observing the color of the soldiers' trimmings, one may easily distinguish to what branch of the service they belong.

DISTINGUISHING COLORS.

The garrison flag is the American flag.

For artillery—the regimental color is yellow.

For infantry—blue.

Cavalry—small guidons, half red and half white—the red uppermost. The flag is swallow-tailed.

Flag of truce—white.

Surgeon's quarters—usually white.

CAMP SIGNALS.

The Reveille—is the signal to rise.

Peas upon a Trencher—call for breakfast.

The Troop—call for guard and guard mounting.

Roast Beef—the call for dinner.

The Assembly—to form companies.

The Retreat—call for duty, and reading orders.

The Tattoo—the signal for retiring, and lights put out, after which none are allowed to leave the camp unless by special permission of the officer of the guard.

OFFICERS' BAGGAGE.

When troops are moving, officers' baggage is not allowed to exceed (mess chest and all personal effects included) as follows:

General officers', 125 pounds.

Field officers', 100 pounds.

Captains', and other officers of the same and under rank, 80 pounds.

PRECAUTIONS.

If arms are taken apart to clean, it must be done by detachments successively, to guard against surprise.

—The number killed in war is small compared to the large numbers carried off by disease, exposure, and other casualties incident to war, which may be guarded against by very simple precautions.

—Soups in the morning are preferable to coffee, and for this purpose the meat should be kept slowly stewing over night.

—When an army is not moving, rations are issued for four days at a time, commencing with companies to the right of a regiment, and finishing with the left company.

—Sentinels will fire on all persons deserting to the enemy.

—In detached corps, small posts of picked men are at night sent forward to watch the enemy and give signal of his approach.

—Sentinels should be relieved every two hours, to keep them on the alert while on duty.

—With raw troops, or when the night is stormy and dark, sentinels should be doubled.

—A sentinel must be sure of the presence of an enemy before he fires; once satisfied of that, he must fire, though all defence on his part is useless. The safety of the post depends upon his vigilance.

—General officers appoint their own *aids-de-camp.*

—Every soldier should have a company number.

—Deserters in time of war, if caught, are punished with death.

—One sutler is allowed for every regiment.

—Escorts of honor should be composed of cavalry or infantry, or both.

—In case of disorder a sentinel must call out the guard.

—Inexperienced officers are put on guard as supernumeraries, for the purpose of instruction.

—General orders of commanders are numbered.

—In signing an official communication, the writer should annex to his name his rank and corps.

—Official letters should refer to one matter only.

—All communications on official service, should be marked on the cover "official business."

—The formation by divisions is the basis of the organization and administration of armies in the field.

—Mixed brigades are often formed of infantry and light cavalry. They are used for advanced guards.

—Heavy cavalry belongs to the reserve.

—In military operations, brigades and divisions are designated by the officers commanding them.

—The grand depots of an army are established where the military operations would not expose them to be broken up.

—Reconnoissances should precede the establishment or the camp.

—The camping party of a regiment consists of the regimental quartermaster and quartermaster sergeant, a corporal and two men per company.

—Fatigue parties are usually sent for supplies.

—The tents are arranged in ranks and files.

—The messes of prisoners should be sent to them by the cooks.

—*File left!* is the command to wheel to the left, and *file right!* the command to the right.

—Sergeants, like the rank and file, will always, under arms, appear with bayonets fixed.

—Pioneers and drummers may be designated as markers, and used accordingly in the manoeuvres and evolutions.

—Before forming a company, drill the men in squads.

—You can see a shell come toward you, but not a cannon ball; but when you fire a cannon ball, you can see it go from you.

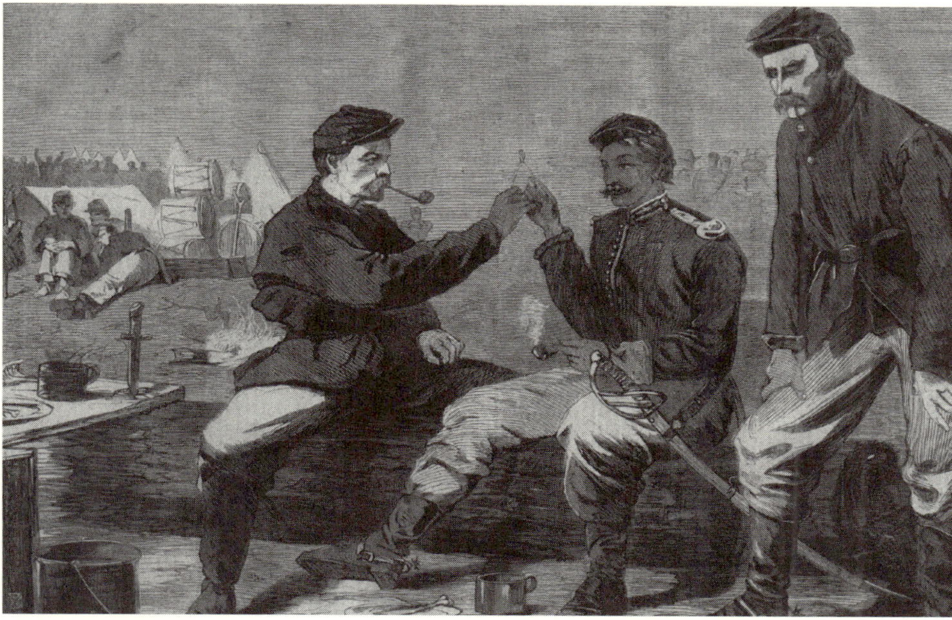

"Civil War Cavalrymen split a turkey bone, Thanksgiving, 1864."
Source: Winslow Homer Civil War Prints.

COOKING, &C.

TO PREPARE YEAST.

One handful hops, boiled half an hour in two quarts water.

10 potatoes, boiled half an hour and mashed; strain the water from the hops on the potatoes, sit in 2 tablespoons of salt and 1 pint flour; set it to cool, add 1 pint brewers' yeast and let it rise 6 hours; strain and put into a tight stone jug.

TO MAKE BREAD.

3 quarts flour.
1 half cup yeast.

1 tablespoon of salt, and warm water enough to make dough; knead till smooth; get the flour off the sides of the pan, set in a warm place to rise over night; in the forenoon knead the dough, divide it, put it into pans, leave it to rise an hour, and bake it three-quarters of an hour.

MIXED BREAD.

1 part rye meal.
1 part Indian meal, scalded.
1 part wheat flour.
Hal cup yeast; mix and set to rise eight hours; bake hot 40 minutes.

BISCUIT.

2 quarts flour.
2 ounces butter.
Half pint boiling water.
1 teaspoon salt.
1 pint cold milk.
Half cup yeast.
Mix well and set to rise, then mix a teaspoonful of saleratus in a little water and mix into the dough, roll on a board an inch thick, cut into small biscuits and bake 20 minutes.

SLAPJACKS.

Take flour, little sugar and water, mix with or without little yeast, the latter better if at hand, mix into paste, and fry the same as fritters in clean fat or butter.

SOUR MILK BISCUIT.

1 quart flour, 1 pint sour milk, 1 teaspoonful saleratus. Mix in the milk till it froths, stir it into the flour cold, mix quick and bake in the oven.

A SIMPLE SOUP.

Skim fat from mutton or beef steak, put into a pot with two or three carrots, turnips and onions, a cup of rice, the bones and bits of cold meat, pepper and salt; boil four hours, and then take out the bones.

COLD BEEF, BONES, BITS OF MEAT.

Put them in a pot with 4 quarts of water, 2 or 3 carrots, turnips, onions, a few cloves, pepper and salt, boil four hours, strain and put back into the pot, mix a tablespoon of flour with water. Stir it in and boil ten minutes.

USEFUL HINTS.

Drippings of beef and pork are nice to fry in, clarified. To clarify, put the drippings into an iron pot; to a quart of fat allow half a pint of cold water; let it boil till the water is boiled out, which may be known by its not bubbling. Strain it, and it is ready for use.

ROAST BEEF.

Roast ten minutes to every pound; don't salt or flour it till nearly done.

BEEF GRAVY.

Take drippings from the meat, put into a pan, and add a cup of boiling water, shake in a little flour and salt. Let it come to a boil, stirring it always.

A LEG OF MUTTON.

Should be roasted ten minutes to the pound. A shoulder the same, except half an hour less time to boil.

A BREAST OF VEAL.

Takes one and a half hours to roast. Lay a few strips of pork across. Stuff with bread, onions and potatoes, chopped fine and seasoned.

TO BOIL RICE.

Wash clean two cupfuls, put it into a pot with two quarts of water, and boil it quite tender; drain, but do not stir it. Let it dry before the fire ten minutes. Every kernel will be separate, dry, and look white.

TO MAKE COFFEE.

Dry it in the oven, with the door open, one or two hours before roasting. When ready, set it on the fire, and stir till a light brown. One cup of ground will make one quart. Ten minutes are long enough to make good coffee.

TO MAKE TEA.

Put it in the tea while the pot is hot; turn in just water enough to *wet* the tea, and let it stand five minutes; then fill it up with boiling water.

BREAD PUDDING.

Take a pound of stale bread; boil a quart of milk, pour it on and let it soak two hours; then rub it fine with the hands. Beat up five eggs and add them; and also a tablespoon of any kind of spice, two cups of sugar, and a little butter. Bake or boil two hours.

RICE PUDDING.

Mix three large spoonfuls of rice in a little cold milk; stir it into a quart of boiling milk; let it boil fifteen minutes. When cold, add a little lemon, sugar, and bake one hour.

Part Sixth.

ELEMENTARY PRINCIPLES OF THE MANUAL—FORMATION OF COMPANY REGIMENT.

POSITION OF THE RECRUIT.

At the command ATTENTION! stand perfectly square, head erect, eyes direct to the front, arms allowed to fall naturally by the side, the thumb and two fingers turned slightly outwards; heels together, and toes four inches apart, knees straight, and body inclining a little forward.

The recruit should learn to assume this position with ease, and without stiffness.

***Forward*—MARCH!**

At the command MARCH! he should invariably step with the left foot first, twenty-eight inches each step, considered a pace, without swaying the body or swinging the arms.

HALT!

At the command HALT! immediately draw up the right foot to the left, and come to a halt with the body erect, as in first position.

***Eyes*—RIGHT!**

At the word RIGHT! turn the head gently to the right, and fix your eyes on a line with the eyes of those next to you in the same rank.

Front! Assume your former position, with eyes direct to the front.

***Eyes*—LEFT!** The same movement, to the left.

Care should be taken not to swerve or move the body at any of these commands.

REST!

Draw the right foot a few inches to the rear of the left, leaning the weight of the body on the right foot, and cross the palm of the left hand over the back of the right hand, in front.

FACINGS!

Facing to the right is executed by placing the hollow of the right foot within an inch of the heel of the left foot. At the command *Right*—FACE! the recruit will wheel to the right, and then assume an erect position, drawing the heels close together.

***Left*—FACE!** A reverse movement, to the left.

***About*—FACE!**

At the word *About!* the left toe should be placed within an inch of the right heel; and at the second command, FACE! turn the body to the left on the heels, and face to the rear, bringing the heels again close together.

COMMON TIME.

Common time in marching is at the rate of ninety steps per minute.

DIRECT STEP.

The direct step is twenty-eight inches.

DOUBLE QUICK STEP.

The double quick step is thirty-three inches, and is made at the rate of one hundred and sixty-five steps per minute.

***Double quick step*—MARCH!**

At the first command, the recruit will raise his hands closed, the fingers towards the body, to a level with the hips; the elbows to the rear, working free. At the second command, raise the left leg, the toe depressed, keeping step with those in the rank, and continue the movement till the command.

HALT!

Immediately stop, drop the hands by the side, draw the heels together, and assume first position.

RUNNING.

The principles are the same as in double-quick step. Care should be taken to preserve the rank unbroken. Room should be afforded for free use of the limbs, without those in the rear, or to the right or left, pushing or crowding.

The breath may be recovered when the recruit is out of wind, by elevating the arms behind the head occasionally. Relief may also be had by throwing the head upwards at times.

A SQUAD.

A squad consists of three or twelve men.

A RANK

is a straight line of a platoon, company, or battalion.

A FILE.

A line of men, one posted behind another.

A PLATOON.

Half a company.

A SECTION.

is half of a platoon.

A COMPANY.

consists of seventy-four men, according to army regulations. Company officers and sergeants are nine in number.

A REGIMENT.

Ten companies compose a regiment, or seven hundred and forty men. Each company is named in alphabetical order—viz: Co. A, B, C, D, E, F, G, H, J, K. The whole commanded by a colonel.

A BATTALION.

is half a regiment, or two or five companies.

A BRIGADE.

Two regiments or more compose a brigade, commanded by a brigadier-general.

A DIVISION.

Two brigades to a division, commanded by a major-general.

A CORPS D'ARMEE.

One or more divisions compose a *corps d'armée*, commanded by a major-general, or a lieutenant-general.

FORMATION OF A COMPANY.

The company being assembled on its ground, the sergeants, by the command *Fall in!* cause the rank and file (corporals and privates) to form in one rank, faced to the *right*, and in the order of height from right to left. (head of the rank to the foot;) the tallest man on the right, (now head of the rank,) the next tallest immediately convering the first, and so on to the left, or rear of the rank.

The first sergeant then gives the word.

Front!—FACE! At the word FACE! the company faces to the front, and the second sergeant places himself on the left of the rank.

The order is then given to *count off.*

The men then count off from the right, as *one, two;* the third man counts *one* again, and the fourth man *two,* and so on to the left of the rank.

The number two men compose the rear rank, and number ones the front rank.

In two ranks form company! At this command number twos will take one step to the rear, number ones standing fast.

Close—RANKS! At the word RANKS! the men close together. Number *two* are then ordered to take on step to the right, which they do by stepping directly behind number one, leaving a space of thirteen inches between their breast and the back of the man in advance. Thus we have the company formed. The captain now divides the company into two equal parts. The first, composing the largest men, is called the *first platoon*, and the next, with the smallest men, the *second platoon.*

Each platoon is afterwards divided into two equal sections. They are numbered from right to left—*first, second, third,* and *fourth* section. The corporals (four in number) are then placed on the right and left of each platoon, in the front rank, according to height.

Between each platoon sufficient space should be allowed for wheeling to the right or left.

The officers and sergeants now take their posts as follows:

The captain in the front rank, on the right of the company, his left elbow slightly touching the right of the front rank man.

The first lieutenant, two paces in rear of the rear rank, and equi-distant between the centre of the second platoon and the second file from the left of the company.

The second lieutenant opposite the centre of the second platoon, and two paces in rear of the rear rank.

The first sergeant, on the right of the rear rank, covering the captain; he is denominated *covering sergeant,* or *right guide* of the company. The second sergeant, two paces in rear of the second file from the left of the company; he is denominated the *left guide* of the company. In the left company of a battalion, this sergeant is on the left of the front rank; he is designated as the *closing sergeant,* and the corporal, the *covering corporal.*

The third sergeant, two paces in rear of the second file from the left of the first platoon.

The fourth sergeant, two paces in rear of the second file from the right of the second platoon.

The fifth sergeant, two paces in rear of the rear rank, and equi-distant between the second lieutenant and the third sergeant.

The pioneer is posted in the line of file-closers on the right; and the music in a line with the front rank, four paces on its right; the drummer on the right of the fifer or bugler. Absent officers and sergeants may be replaced, officers by sergeants, and sergeants by corporals, according to rank and the necessity of the case.

TO PASS FROM TWO RANKS INTO ONE RANK.

The order is given, *In one rank, form company!* The left guide faces to the left.

MARCH! The left guide steps off and marches in the prolongation of the front rank; the next file steps off at the same time with the guide; the front rank man turns to the left at the first step, follows the guide, and is himself followed by the rear rank man of his file, who turns on the spot where his file-leader turned. The second file, counting from the left, and successively all the other files, march as prescribed for the first; the front rank man of each immediately following the rear rank man of the file next on the left. The captain *halts* the company as the last man on the right turns into the rank. The file-closers extend themselves with the movement.

IN COLUMN BY COMPANY.

The captain is two paces in front of the centre of his command; the first sergeant on the right of the front rank, and is the right guide of the company; the second sergeant on the left of the same rank, and is the left guide of the company.

IN COLUMN BY PLATOON.

The captain commands the first platoon, the first lieutenant the second platoon; each takes post two paces in front of the centre of his platoon; the first sergeant is the guide of the first platoon, and the second sergeant is the guide of the second platoon. They are on the left of the front rank of their respective platoons, if the column be right on front; and on the right (the company is reversed) if the left be in front.

If the column be marching in the route step, the chiefs of platoon take the place of the guides on the directing flank, and are covered by the latter in the rear rank.

IN COLUMN BY SECTION.

This column takes place only in column of route, and then only when platoons have a front of ten or more files. The captain commands the first section, the first lieutenant the third, the second lieutenant the second, and the third lieutenant the fourth section; —each taking post on the directing flank in the front rank of his section; the two guides who are thus displaced, each falls back to the rear rank of his section, and covers its chief; the remaining file-closers place themselves in the rear rank of their respective sections, all on the side of direction.

In column, except in column by sections, the file-closers, not otherwise provided for, are in their proper places behind the rear rank of their respective subdivisions.

In close column, or in column at half distance, they close up to within one pace of the rear rank. In column right in front, the left flank is the directing flank, (except under peculiar circumstances or manoeuvres.) The reverse is the case in column left in front.

TO OPEN RANKS.

The company being at shoulder arms, the left guide is placed on the left of the rear rank.

To the rear—Open order!—At the word of command, the covering sergeant and the left guide step off to the rear, in the back step, four paces, and allign themselves parallel to the rear rank.

MARCH!—The front rank stands fast, the rear steps off in the back step, in common time, without counting steps, placing itself on the allignment marked out for it and is alligned by the right on the left guide, by the covering sergeant. The file-closers step off at the same time with the rear rank, and place themselves two paces in the rear of that rank.

The ranks being properly alligned, the command *Front!* is given.

ALLIGNMENTS IN OPEN RANKS.

The ranks are the first alligned man by man, as in the squad drill, three men being placed three paces in advance of the right or left of each rank, to serve as a basis, and the command given, *By file right* (or left)—*Dress!*—The men of each rank move up successively on the allignment, each man being preceded two paces by his neighbor. The ranks are next alligned at once, forward and backward, in both parallel and oblique directions, by the commands *right* (or left) *dress!* or *right* (or left) *backward dress!* three men, in each case, being placed on a basis. Before closing the ranks, the company should be exercised in the manual of arms, and the loadings in nine and four times and at will.

TO CLOSE RANKS.

Close order—March!—At the command MARCH! the rear rank closes to the front, the men in the rear rank stepping briskly up, and covering his file-leader.

The company should next be instructed in the allignments and the manual of arms in closed ranks, and then proceed to the loadings and firings.

TO FIRE BY COMPANY.

Fire by company!—At this command the captain places himself opposite to the centre of his company, four paces in rear of the rank of file-closers. The covering sergeant places himself in that rank, opposite to his interval.

This rule is general for the captain and covering sergeant in all the different firings.

Commence—Firing!—At this command, the captain gives the words *company—ready—aim—fire—load!* At the word *load* each man brings back his piece, loads, and comes to a shoulder, when the captain recommences the fire by the same commands, and thus continues it till orders to *cease firing,* (or the roll of the drum,) from his superior, (say colonel, if the company is attached to a regiment, and is drawn up in line of battle).

The captain may sometimes cause aim to be taken to the right and left, by giving the word *right* (or left) *oblique* between the words *ready* and *aim.*

TO FIRE BY FILE.

Fire by file—Company—Ready—Commence—Firing—At the words *Commence Firing!* The right file aims and fires, and the next file aims at the instant the first has fired, and so on successively. After the first fire every man loads and fires without waiting for the others. (Fire at will.) Each man faces to the front in casting about, and after returning rammer springs up his piece with his left hand, and at the height of his chin, and makes a half face to the right, taking the position of *Ready.*

TO FIRE BY THE REAR RANK.

Face by the rear rank!—At this command the captain steps out and places himself near to, and facing the right file of the company; the covering sergeant and file-closers pass quickly through the captain's interval, and place themselves faced to the rear, and covering sergeant a pace behind the captain, and the file-closers two paces from the front rank, opposite to their places in line.

Company about—face!—At the word *face!* given the instant the last file-closer has passed through the interval, the company faces about; the captain places himself in the interval in the rear rank, (now front,) and the covering sergeant covers him in the front rank (now rear.) The different firings are now executed in the manner already prescribed.

TO RESUME THE PROPER RANK.

Face by the front rank!—This is executed as prescribed in the command, *face by the rear rank—Company—About—* FACE! The company having faced about, the captain and covering sergeant resume their places in line.

FORMATIONS OF A REGIMENT IN ORDER OF BATTLE.

The regiment is supposed to consist of ten companies, to wit: one grenadier company, one of light infantry or rifles, and eight battalion companies. Since the introduction of the famous zouave movements under the French military system, and the application of modern appliances of warfare, nearly all our volunteer regiments organize as rifle, or light infantry regiments, instead of as heavy infantry of the line. Opinions of military men differ, however, as to which is the most effective in actual combat. In our opinion, heavy infantry of the line belongs more properly to a past age, and could scarcely be near as efficient as a zouave or rifle regiment, well skilled in the sabre bayonet exercise, and the tactics so recently adopted by the French and United States governments. The modern system seems to meet with more favor, as it certainly should among our citizen-soldiers. It affords quicker movements and combinations being made on the field, and therefore is not only really more serviceable in time of action, but less liable to loss by frequent shelling from rifled cannon in position at long range, than dense masses and slow-moving bodies of heavy infantry of the line. The latter may repel a charge as well; but in making a charge, or assaulting an intrenched position, the palm of superiority must be awarded to the former.

A regiment in order of battle is posted from right to left as follows: first company, sixth, fourth, ninth, third, eighth, fifth, tenth, seventh and second, according to the rank of captains.

With a less number of companies, the same principle will be observed: the first captain assigned the command of the right company, the second captain the extreme left company of the regiment, the third captain the right centre company, and so on till the command is full. The companies thus posted are designated from right to left—*first* company, *second* company, and so on to the ten companies. The regimental colors should be posted on the left of the right centre battalion company. The five companies to the right of it are called the *right wing* of the battalion, and the five companies to the left *left wing*.

The *color-bearer* should be a sergeant selected by the colonel, with two ranking corporals on his right and left. The guard is composed of six corporals in addition. There should be no display of colors in battalions of less than five companies, and no color guard, except it may be at the reviews.

There are two *general guides* to each battalion, selected from among the second or third sergeants by the colonel. The first one is called the *right general guide,* posted in rear of the right flank of the battalion, and the second the *left general guide* posted in the line of file-closers, in rear of the left flank of the battalion. The field music is posted twelve paces in rear of left centre company, and band five paces in rear of the field music.

FIELD OFFICERS AND REGIMENTAL STAFF.

The field officers, colonel, lieutenant-colonel, and major should be mounted on horseback. The adjutant, when the battalion is manoeuvring, will be on foot.

The colonel's post is thirty paces in the rear, and opposite the centre of the battalion.

The lieutenant-colonel twelve paces in the rear, and opposite the centre of the right wing.

The major twelve paces in the rear, and opposite the centre of the left wing.

The adjutants and sergeant-major opposite the right and left of the battalion, respectively, and eight paces in the rear of the file-closers.

The quartermaster, surgeon, and other staff officers, in one rank, on the left of the colonel and three paces in his rear.

The quartermaster-sergeant, on a line with the front rank of the field music, and two paces on the right.

The adjutant and sergeant-major will aid the colonel and major in the manoeuvres.

In the absence of the colonel, then the lieutenant-colonel will take command. In his absence, the major, and the major's absence then the senior captain.

OFFICERS AND NON-COMMISSIONED OFFICERS.

Officers and non-commissioned officers are in the following rank:

1. Lieutenant-general,
2. Major-general,
3. Brigade-general,
4. Colonel,
5. Lieutenant-colonel,
6. Major,
7. Captain,
8. First Lieutenant,
9. Second Lieutenant,
10. Ensign,
11. Sergeant-major,
12. Quartermaster-sergeant of a regiment,
13. Quartermaster-sergeant, and Hospital Steward,
14. First Sergeant,
15. Sergeant,
16. Corporal.

Each grade takes precedence in rank by date of commission, or appointment. Brevet rank is a mere honorary distinction, and does not give precedence.

Officers commissioned by any State take rank next *after* officers of like grade commissioned by the Federal Government.

Resignations will not take effect until information of their acceptance is made known.

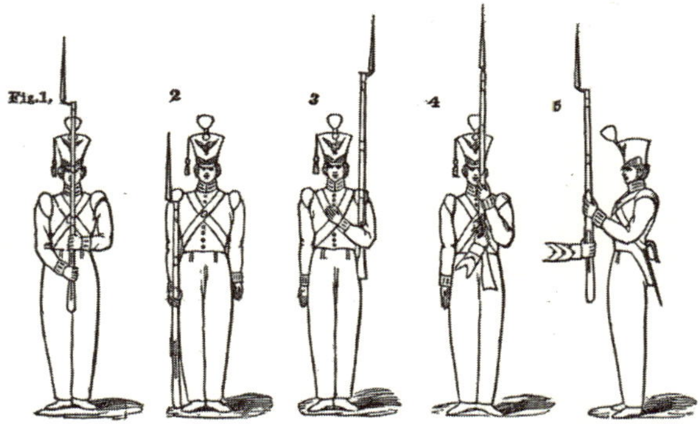

"School of the Soldier—Present Arms and Inspection, 1830." Source: U.S. War Department, *Abstract of Infantry Tactics*, Washington, D.C.: War Department, 1830.

PRINCIPLES OF SHOULDERED ARMS.

The piece in the right hand, the barrel resting in the hollow of the shoulder, the guard to the front, the arm bent slightly, and hanging easy near the body, the thumb and fore-finger embracing the guard, the remaining fingers closed together and grasping the swell of the stock, directly under the cock, which rests on the little finger.

Care will be taken that the piece be not carried too high or too low; the right elbow in the rank should be close to the body, and sufficient space allowed the soldier to handle his piece with facility.

Support—**ARMS!**

(First motion.)—Bring the piece with the right hand to the front, perpendicularly between the eyes, the barrel to the rear;

seize the piece with the left hand at the lower band, raise this hand as high as the chin, and seize the piece at the same time with the right hand four inches below the cock.

(Second motion.)—Turn the piece with the right hand, the barrel to the front; carry the piece to the left shoulder, and pass the fore-arm, extended on the breast between the right hand and the cock; support the cock against the left fore-arm, and then rest the left hand on the right breast.

(Third motion.)—Drop the right hand by the side.

REST!

At this word, bring up quickly the right hand to the handle of the piece, (small of the stock.)

When this position is assumed, it will not be required to preserve silence, or steadiness of position.

When the instructor may wish the recruit to pass from this position to that of steadiness he will command

***Attention*—SQUAD, (or COMPANY!)**

At the second word, the recruit will drop the right hand to the side, and maintain silence in the rank.

***Shoulder*—ARMS!**

One time and three motions.

(First motion.)—Grasp the piece with the right hand under and against the left fore-arm; seize it with the left hand at the lower band, the thumb extended; detach the piece slightly from the shoulder, the left fore-arm along the stock.

(Second motion.)—Carry the piece vertically to the right shoulder with both hands, the rammer to the front; change the position of the right hand so as to embrace the guard with the thumb and fore-finger, slip the left hand to the height of the shoulder, the fingers extended and joined, the right arm nearly straight.

(Third motion.)—Drop the left hand quickly by the side.

***Present*—ARMS!**

One time and two motions.

(First motion.)—With the right hand bring the piece erect before the centre of the body, rammer to the front; at the same time seizes the piece with the left hand half way between the guide-sight and lower band, the thumb extended along the barrel, and against the stock, the fore-arm horizontal, and resting against the body, the hand as high as the elbow.

(Second motion.)—Grasp the small of the stock with the right hand below and against the guard.

***Shoulder*—ARMS!**

One time and two motions.

(1). Bring the piece to the right shoulder, at the same time change the position of the right hand so as to embrace the guard with the thumb and forefinger; slip the left hand to the height of the shoulder, the fingers extended and joined, the right arm nearly straight.

(2). Drop the left hand quickly by the side.

***Order*—ARMS!**

One time and two motions.

(1). Seize the piece promptly at the word *Order!* with the left hand near the upper band, and detach it slightly from the shoulder with the right hand; loosen the right hand, lower the piece with the left; re-seize the piece with the right hand above the lower band, the little finger in rear of the barrel, the butt about four inches from the ground, the right hand supported against the hip; drop the left hand by the side.

(2). Let the piece slip through the right hand, by opening slightly the fingers, to the ground. The hand low, the barrel between the thumb and forefinger, extending along the stock; the other fingers extended and joined; the muzzle about two inches from the right shoulder; the rammer in front; the tow (or beak) of the butt against and in a line with the toe of the right foot; the barrel perpendicular.

REST!

At this command, assume an easy position, standing, of course, in the rank.

***Attention*—SQUAD, (or COMPANY!)**

At the second word, resume the position of order arms.

***Shoulder*—ARMS!**

One time and two motions.

(1.) Raise the piece vertically with the right hand to the height of the right breast, opposite shoulder; elbow close to the body; seize the piece with left hand, below the right, and drop quickly the right hand to grasp the swell of the stock, the thumb and forefinger embracing the guard; press the piece against the shoulder with the left hand; the right arm nearly straight.

(2.) Drop left hand quickly by the side.

***Load in nine time*—LOAD!**

Grasp the piece with the left hand, as high as right elbow; bring it vertically opposite middle of the body; shift the right hand to the upper band; place the butt between the feet, the barrel to the front; seize it with left hand near the muzzle, which should be three inches from the body; carry the right hand to the cartridge-box.

***Handle*—CARTRIDGE!**

Seize cartridge with thumb and two fingers, and place it between the teeth.

***Tear*—CARTRIDGE!**

Bite the end of the paper to the powder; hold the cartridge upright, and in this position place it in front, and near the muzzle; keep the back of the hand to the front.

***Charge*—CARTRIDGE!**

Empty the powder quickly into the barrel; disengage the ball from the paper with the right hand and the thumb, and first two fingers of the left; insert it in the bore, the pointed end uppermost, and press it down with the right thumb; seize the head of the rammer with the thumb and forefinger of the right hand; the other fingers closed, and elbows near the body.

***Draw*—RAMMER!**

Half draw the rammer by extending the right arm; steady it in this position with the left thumb; grasp the rammer with the right hand near the muzzle, the little finger uppermost, the nails to the front, the thumb extended along the rammer.

Clear the rammer by drawing it forth; turn the rammer, the little end passing near the left shoulder; place the head of it on the ball, the back of the hand to the front.

***Ram*—CARTRIDGE!**

Insert the rammer, and press the ball home.

***Return*—RAMMER!**

Draw the rammer half way out with the right hand, and steady it in this position with the left thumb; run the right hand down the rammer again to the muzzle; extend the arm, turn the rammer, the head near the left shoulder; stead it in its "pipes" with the left thumb and forefinger, and force it home; at the same instant, passing the left hand down the barrel to the extent of the arm, and steady your piece without depressing the shoulder.

PRIME!

One time and two motions.

(1.) With the left hand raise the piece till the hand is as high as the eye; grasp the small of the stock with the right hand; half face to the right, at the same time place the hollow of the right foot within an inch of the left heel; slip the left hand down to the lower band, the thumb along the stock; left elbow against the body; bring the piece to the right side, the butt below the right forearm, the small of the stock against the body, and two inches below the right breast, the barrel upwards, and the muzzle on a level with the right eye.

(2.) Cock your piece with the thumb; then take a cap from the pouch, place it on the nipple, and press it down with the thumb; after which, seize the small of the stock with the right hand.

Shoulder—ARMS!

(1.) Bring piece to right shoulder, and support it with left hand.

(2.) Drop left hand quickly to the side.

READY!

Bring the butt to the right shoulder, and lower the muzzle of the piece with both hands; cock and seize the piece at the small of the stock with the right hand, the left extending along the barrel; the muzzle as high as the eye.

Keep the right thumb on the head of the cock, and the other fingers under and against the guard.

AIM!

Cast your eye along the sight of the piece to the object aimed at; carry the right foot twelve inches to the right, inclining the upper part of the body forward; keep the right elbow on a line with the shoulder.

FIRE!

Press the forefinger of the right hand against the trigger, and fire without lowering or turning the head, and remain in this position. If, after the firing, the instructor should not wish the recruits to re-load, he will order shoulder arms.

***Recover*—ARMS!**

To accustom the recruit to wait for the command *fire!* after the order is given to aim, the instructor should command *recover*—arms! At the first part of the command withdraw the finger from the trigger; at the command *arms!* Assume the position of *ready*.

The recruit should be frequently practised in the manoeuvre to *aim* and then *recover arms*, to familiarize him with the word of command.

***Secure*—ARMS!**

(1.) Bring the piece with the right hand to the front, perpendicularly to the front, and between the eyes, the barrel to the rear; seize the piece with the left hand at the lower band, raise this hand as high as the chin, and seize the piece with the right hand at the small of the stock.

(2.) Turn the piece with both hands, the barrel to the front; bring it opposite the left shoulder, the butt against the hip, the left hand at the lower band, the thumb as high as the chin, and extended on the rammer; the piece erect and detached from the shoulder, the left forearm against the piece.

(3.) Reverse the piece, pass it under the left arm, the left hand remaining at the lower band, the thumb on the rammer, to prevent it from sliding out, the little finger resting against the hip, the right hand falling at the same time by the side.

***Arms*—AT WILL!**

Carry the piece at pleasure, on either shoulder, with one or both hands, the muzzle always up.

***Arms*—PORT!**

Throw the piece diagonally across the body, the lock to the front, seize it smartly, at the same instant, with both hands, the right at the handle, the left at the tail band, the two thumbs pointing towards the muzzle, the barrel sloping upwards and crossing opposite to the point of the left shoulder, the butt proportionately lowered. The palm of the right hand above that of the left, under the piece, and the nails of both next to the body, to which the elbows are closed.

***Carry*—ARMS!**

(1.) Carry quickly the right hand to the small.

(2.) Place quickly the left hand on the butt.

(3.) Let fall smartly the right hand into its position, and drop with the left, at the same time, the piece into the position of shoulder arms.

RIGHT SHOULDER SHIFT!

(1.) Supposing the recruit was at a shoulder arms; detach the piece perpendicularly with the right hand, and seize it with the left between the lower band and guide-sight, raise the piece, the left hand at the height of the shoulder and four inches from it; place at the same time the right hand on the butt, the beak between the first two fingers, the other two fingers under the butt plate.

(2.) Quit the piece with the left hand, raise and place the piece on the right shoulder with the right hand, the lock plate upwards.

Let fall, at the same time, the left hand by the side.

***Ground*—ARMS!**

(1.) From a shoulder arms.—Turn the piece with the right hand, the barrel to the left; at the same time seize the cartridge-box with the left hand, bend the body, advance the left foot, the heel opposite the lower band, the knees slightly bent, and lay the pieces on the ground with the right hand.

(2.) Rise up and bring the feet together, and drop the hands by the side.

***Raise*—ARMS!**

(1.) Seize the cartridge-box with the left hand, bend the body, advance the left foot opposite the lower band, and seize the piece with the right hand.

(2.) Raise the piece, bring the feet together, turn the piece with the right hand, the rammer to the front, at the same time drop the left hand by the side.

***Fix*—BAYONET.**

From a shoulder arms.

(1.) Grasp the piece with the left hand at the height of the shoulder.

(2.) Quit the piece with the right hand, lower it with the left hand opposite the middle of the body and place the butt between the feet, the rammer to the rear, the barrel vertical, the muzzle three inches from the body; seize it with the right hand at the upper band, and carry the left hand to the bayonet.

(3.) Draw the bayonet, and fix it on the end of the barrel, seize the piece with the left hand, the arm extended, the right hand at the upper band; come to a shoulder arms.

***Charge*—BAYONET.**

(1.) Raise the piece slightly with the right hand, and make a half face to the right, place the hollow of the right foot opposite to, and three inches from the left heel, the feet square, seize the piece at the same time with the left hand a little above the lower band.

(2.) Bring down the piece with both hands, the barrel uppermost, the left elbow against the body; seize the small of the stock at the same time with the right hand, which will be supported against the hip, the point of the bayonet as high as the eye.

***Trail*—ARMS!**

From a shoulder arms.

(1.) The same as the first motion of order arms.

(2.) Incline the muzzle slightly to the front, the butt to the rear, and about four inches from the ground. The right hand, supported at the hip, will so hold the piece that the rear rank men may not touch, with their bayonets, the men in the front rank.

***Unfix*—BAYONET!**

From a shoulder arms.

(First and second motions.) The same as the first and second motions to fix bayonet, except that the thumb of the right hand will be pressed on the spring, if a sabre bayonet, and with the same hand wrest the bayonet off the edge (or point) to the front, and return it to its scabbard, and seize the piece with the left hand, the arm extended. The next movement, shoulder arms.

To *Stack*—ARMS!

The squad being in two ranks at order arms.

Stack—ARMS!—At the command, the front rank man of every even-numbered file passes his piece before him, seizing it with the left hand above the middle band, and places the butt behind and near the right foot of the man next on the left, the barrel turned to the front. At the same time the front rank man of every odd-numbered file passes his piece before him, seizing it with the left hand below the middle band, and hands it to the man next on the left; the latter receives it with the right hand two inches above the middle band, throws the butt about thirty-two inches to the front, opposite to his right shoulder, inclining the muzzle towards him, and locks the shanks of the two bayonets, the lock of this second piece towards the right, and its shank above that of the first piece.

The rear rank man of every even file projects his bayonet forward, and introduces it (using both hands) between and *under* the shanks of the two other bayonets. He then abandons his piece to his file-leader, who receives it with the right hand under the middle band, brings the butt to the front, holding up his own piece and the stack with the left hand, and places the butt of this third piece between the feet of the man next on the right.

The stack thus formed, the rear rank man of every odd file passes his piece into his left hand, the barrel turned to the front, and sloping the bayonet forward rests it on the stack.

Break—RANKS.

To dispose of the command, the order is given to—*Break ranks*—MARCH!

To resume—ARMS!

The squad, on a signal, or order to fall in, re-forms in two ranks.

The next order is to resume arms, when each man advances and seizes his piece.

When companies stack arms, the sergeants, and also corporals, if in the rank of file closers, rest their pieces against the stacks nearest to them, respectively, *after* ranks are broken, and resume their pieces on the signal to reform ranks.

The color-guard forms a separate stack.

LOAD AND FIRE AT WILL!

When the command is given to load and fire at will, each recruit will load and fire with care, but at the same time with celerity and dispatch. Much depends on beating an enemy to load and fire two or three times, if possible, to his once. The recruit, therefore, should learn to handle his piece with ease and dexterity.

MARCHING AND WHEELING MOVEMENTS

With the limited space at command, it would be impossible to give full and complete directions of the marching and wheeling movement at present in vogue in the Army of the United States. A competent drill-master can more readily instruct a body of recruits in evolutions and manoeuvres than for them to endeavor to learn the various movements from the long written descriptions, and series of illustrated plates.

We must content ourselves, therefore, in giving the commands only which is absolutely necessary for a recruit to perfect himself in to become a good soldier, without going into extended details. They are to—

Mark time—March.

Forward—March.

To change step.

To march backwards.

To march to the front.

Right oblique.

Left oblique.

To march to the front in double-quick time.

onist. The meeting near the edge of the cliff, apparently so accidental, had been anticipated a week previously by Downey at least, for he had notified me at the time when, where, and under what circumstances it would take place, and what would be the result.

It was precisely as he had predicted.

They met; and as soon as they recognized each other, although Langdon was disguised as an Indian warrior, they discharged their rifles with skillful effect into each others bodies; then threw them aside and closed with their knives, and in the violent struggle which ensued for the mastery, both were lost over the frowning precipice, exactly in the manner as he related to me would be the case. The fate of each was sealed, but one at least, if not both, had been forewarned of the frightful ordeal through which he had just passed.

As the twilight of that lovely summer evening began to wane, and the gathering shades of night to fleck the sward, all that was mortal of Joseph Downey, were consigned to their last and final resting-place adown the valley of the Yo Semite, in the wilds of California.

In twenty days more of successful campaigning we retraced our steps homeward.

PERILS OF THE CALIFORNIA PIONEERS.

"Perils of California Pioneers from *Gazlay's Pacific Monthly*, June 1865."
Source: University of Washington Library MF A231, 265.

THE ATTACK.

(above) "Panning for gold on the Mokelumne River, Calaveras County, CA."
Source: http://wikipedia.org/wiki/Gold_Rush_(California).

(below) "Attack by Native Americans on Miners' Settlement."
Source: http://wikipedia.org/wiki/Gold_Rush_(California).

BIBLIOGRAPHY

Agthe, Claire Keenan. "#6239 John P. Curry-Undertaker's Certificate, 1909." Philadelphia Death Records Online. http://labs.familysearch.org, accessed 18 June 2007.

Coraluzzi, A.S. "Funeral Notices." *Philadelphia Public Ledger,* 24 September 1850, 2.

Curry, John Penn. "An Adventure in the Wilds of California." *Gazlay's Pacific Monthly*, June 1865.

———. "Dottings on Foreign Coasts." *Gazlay's Pacific Monthly*, June 1865.

———. "Musical Warblings—from the Note-Book of a Strolling Amateur." *Gazlay's Pacific Monthly*, June 1865.

———. *Observations and Experiences among the Mineral Regions of Nevada.* New York: William Quinn, 1865. Yale University's Beinecke Library, New Haven, Connecticut, Collection of Western Americana: Frontier History of the Trans-Mississippi West, 1550–1900, no. 1457.]

———. *Volunteers' Camp and Field Book.* Richmond VA: West and Johnson, 1862.

Garrison, Webb. *A Treasury of Civil War Tales.* New York: Ballantine Books, 1993.

Gazlay, David M. *Gazlay's Pacific Monthly.* New York: David M. Gazlay, 1865.

Hallenbeck, Louis and Coreen. Personal letter. 14 July 2007.

Hallenbeck, Louis and Coreen. Genealogy research. New York City Directory 1881/1882, The New York State Library, Cultural Education Center, Albany NY 12230.

Haskins, Charles W. *The Argonauts of California.* New York: Fords, Howard and Hulbert, 1890.

Jones, J. William. *Christ in the Camp.* Richmond VA: Martin and Hoyt Company, 1904.

Lord, Francis A. *Army and Navy Textbooks and Manuals used by the North during the Civil War-Part II.* Volume 9 of *Company of Military Historians, Military Collector and Historian* Carlisle PA: US Army Heritage and Education Center, 1957.

Linda Lorda and Louis J. Rasmussen. *San Francisco Ship Passenger Lists* (1850–1851), vol. 2 (Provo UT: The Generations Network, Inc., 2006).

Lyman, George D. *The Saga of the Comstock Lode: Boom Days in Virginia City.* New York: Charles Scribner's Sons, 1934.

Maryniak, Benedict, "Letter to the editor, 13 March 2007." Ohio Historical Society's Civil War Documents Searchable Database, series 147-6:187. http://ohiohistory.org/CivilWarGuideProject, accessed 13 March 2007.

Mershon, S.L. *The Symes Foundation: A Corporation of Christian Stewardship.* Montclair, NJ: The Symes Foundation, 1923.

Ohio Historical Society. "September 3, 1861," in the Civil War Documents Searchable Database, Series 147-6: 187. http://ohiohistory.org/CivilWarGuideProject, accessed 13 March 2007.

Posehn, Margaret. "Departure." *Daily Alta,* 21 May 1860, page 2, column 1.

Rasmussen, Louis J. *San Francisco Ship Passenger Lists* (1851-1852), vol. 3 (Provo, UT: The Generations Network, Inc., 2006).

Rainey, Lee. "Dining Out with Hood's Texans," *4th Texas Infantry*, http://members.aol.com/h4texas/dining4.htm/, accessed 2 November 2006.

Shaffer, Jeffrey P. *Yosemite National Park.* Berkeley, CA: Wilderness Press, 1999.

US Army. *Revised United States Army Regulations of 1861.* Washington, DC: Government Printing Office, 1863.

US Senate. *Report of the Superintendent of the U.S. Coast and Geodetic Survey, 45th Congress, 3rd session, June 1878.* Washington, DC: Government Printing Office, 1879.

US War Department. *Abstract of Infantry Tactics.* Washington, DC: War Department, 1830.

"Visit of the United States Sloop-of-War St. Mary's in the Navigator Islands." *The New York Times*, 20 November 1856.

Wakefield, Adine. "David M. Gazlay 1858-1875," http://genforum.genealogy.com/gazlay/messages, accessed 19 May 2009.

War Department. *Revised United States Army Regulations of 1861.* Washington, DC Government Printing Office, 1863.

INDEX TO PROPER NAMES